RESEARCH REPORT

Minimum Invasive Fire Detection for Protection of Heritage

Author

Geir Jensen, COWI AS, Norway

Jointly Published by

Riksantikvaren the Norwegian Directorate for Cultural Heritage

Historic Scotland: Technical Conservation, Research and Education Group

in Support of

COST – the European CO-operation in the field of Scientific and Technical Research – Action C17 Built Heritage: Fire Loss to Historic Buildings

ISBN 82-7574-040-1

CONTENTS

ACRONYMS

AAFD	Automatic area fire detection
ASD	Aspirating smoke detector
ASDHS	Aspirating smoke detector, high sensitivity
ASP	Aspirating smoke detector
BFPSA	The British Fire Protection Systems Association
BSD	Beam smoke detector
CACFOA	The Chief and Assistant Chief Fire Officers' Association
CW	Compact wireless all-in-one detector
EN	European Norm
FD	Flame detector
G-JET	Smoke detection design tool
HDP	Point heat detector
HS	Historic Scotland
KP	Compact wireless all-in-one detector
LHD Pair	Line heat detector, short-circuiting type
LHD Tube	Line heat detector, pneumatic type
LHD Wire	Line heat detector, melting wire type
LHD	Line heat detector
LO	Beam smoke detector
NFPA	National Fire Protection Association
NIST	National Institute for Science and Technology
ODPM	UK Government Office of the Deputy Prime Minister
ORD	Point smoke detector, optical
OSDP	Point smoke detector, optical
RNDCH	Riksantikvaren: Norwegian Directorate for Cultural Heritage
RTI	Rate of Temperature Index
SFPE	Society of Fire Protection Engineers
SK	Visual imaging fire detector
SV	Sound and vibration fire detector
SVFD	Sound and vibration fire detector
TIFD	Thermal imaging fire detector
TRD	Aspirating smoke detector, high sensitivity
VD	Point heat detector
VDL Pair	Line heat detector, short-circuiting type
VDL Tube	Line heat detector, pneumatic type
VDL Wire	Line heat detector, melting wire type
VDL	Line heat detector
VIFD	Visual imaging fire detector
VK	Thermal imaging fire detector

1   EXECUTIVE SUMMARY

Fire detection systems in general are effective fire safety measures for heritage buildings and museums. Still, we are faced with these challenges of detectors and inherent cable installations:

- Irreversibly impair fabric or décor
- Renovation and maintenance incur irreversible damage to fabric or décor
- Aesthetically invasive measures in sensitive environments
- Detectors do not respond to fires as quickly as anticipated
- Excessive nuisance alarms: detectors disconnected, or downgraded response
- Cable installations increase risk of fire from lightning
- Application may be inappropriate in terms of cost, efficiency, obtrusiveness

A summary of technologies used for minimizing invasive detector installations has been made. Results are evaluated and recommendations given. The solutions and recommendations generally apply world-wide. Some tests were made to find optimum solutions for the highly valued stave churches of Norway. Other projects and tests were made to investigate outdoor area fire surveillance of historic town centres and multiple building heritage sites.

For indoor applications in historical buildings and museums aspirating smoke detectors are found to be the best option overall to minimize invasion, reversibility, early detection, reliability and on several other factors evaluated. Where heat detection is sufficient, line heat detectors are by far most suited to heritage. Line heat detectors may hardly be visible, are sensitive along their lengths, very reliable, cost-effective and some may be repaired locally with no special parts.

Wireless point detectors are a valued solution to avoid unnecessary invasion. These have improved from the need to exchange batteries frequently, except for non-heated areas in cold climates. The expensive products offer high reliability and are unobtrusive, although most wireless units are quite bulky.

Both visual and thermal image fire detectors (camera software fire detection) may be used in large indoor spaces from well hidden locations. The visual category is prone to deception by moving objects and shadows. The thermal ones are very reliable: They discriminate any movements or shade and detect fire by temperatures exceeding set limits only, but are quite expensive for indoor use.

For surveillance of historic town centres, multiple building heritage sites or all wooden structures externally, thermal imaging fire detectors (cameras) are the most efficient, also because they transmit video to manned alarm stations. Staff will then discriminate harmless incidents from real fires, and avoid unnecessary intervention. Cameras make invasive installations in the buildings unnecessary.

Aspirating smoke detectors are efficient in detecting outdoor or neighbourhood fires even when all sampling tubing and holes are inside a building, and with a low probability of false alarms. This is due to an integrating effect: Small samples of low density smoke in several sampling points will raise an alarm, while quite dense smoke in one sampling point only will not. Aspirating smoke detectors must be located so that their fans do not cause noise problems in churches. Installation must be done carefully to avoid condensation problems in non- or partly heated buildings. In harsher environments dust filter renewal may be frequent. Aspirating smoke detectors are known for unobtrusive installations and considered to be the least invasive detectors for heritage buildings in general. They offer several other benefits, such as the robustness against false alarms while still being very sensitive to real smoke from fire.

Line heat detectors are also reliable and inexpensive. They are valued for their superior reliability in order to activate extinguishing systems. Smoke detectors are required for early detection of incipient fires, but are less reliable to activate systems that may cause secondary damage. Line heat detectors respond earlier than point heat detectors, and typically as fire heat release exceeds 200 kW - integrating types respond as quickly as point smoke detectors to flaming fires.

Evaluations of minimum invasiveness relate to wired point heat or point smoke detectors as references. Point smoke detectors respond to flaming fire as the smoke layer reaches temperatures within 10-15°C above that of normal air which are common values applied in engineering models. Point optical smoke detectors respond to smouldering fires without substantial room air temperature increase, while ionization smoke detectors may not respond at all. Point heat detectors typically respond as the smoke layer reaches 80-300°C depending on the temperature growth gradient. Sprinklers activate typically from 500 kW rate of heat release at normal room heights.

Although the prime consideration of this report is minimum invasive solutions, other suitability issues like early detection, reliability, cost and other factors are evaluated. Engineering principles for optimum application designs, and measures to reduce false alarms, are also included.

2   CONCLUSIONS & RECOMMENDATIONS

2.1   Fire detectors in aesthetically sensitive rooms

Point heat or smoke detectors – the most well known and most widely used detectors – constitute our reference for rating less invasive detector applications. Although, point detectors may in some instances be fairly unobtrusive if applied in ceiling beam shades or painted. Below follow description of the categories and solutions which is found to be most successful as less invasive alternatives.

Aspirating smoke detectors

Aspirating smoke detectors provide the best overall performances, and offer the best potential for minimum invasion and reversible installation in sensitive environments. One should design aspirating detectors by engineering principles (see chapter 3.3.5) to get them the least obtrusive and performing optimally.

Aspirating detectors may be located in adjacent buildings or suitable rooms, so that sampling pipes only run in protected areas. Tiny capillary tubes may be inserted from voids into room. Aspirating smoke detectors are very reliable in harsh environments.

In historical buildings or galleries, where sampling pipes may not be hidden in voids, the pipes may be made small by careful engineering in copper or other thin-walled tubing. Small wireless point smoke detectors or beam smoke detectors may be considered, but offer many more challenges.

Heat detectors

Most line heat detectors offer these benefits to heritage:

- Detection performance equivalent to one point detector at every point of line
- Very reliable
- Hardly visible at ceiling level
- May be run in ceiling voids with 2 mm diameter 50 mm drops only visible in rooms below
- May be run unobtrusively along cracks or shadowed areas
- May be fed into inaccessible voids via existing tiny holes or cracks
- Very robust
- Low cost
- Low cost maintenance

Following a series of realistic fires[15] a professional guesstimate concluded that melting wire line heat detectors activate before fire output exceeds 200 kW both for indoor and outdoor scenarios when properly installed.

One might also consider wireless point smoke or thermal or visual imaging detectors.

Minimum invasive techniques

To summarize, in most cases the following installations can be considered for minimum invasive detection:

- Linear heat detection
- Aspirating smoke detectors
- Wireless heat/smoke detectors
- Thermal or visual smoke detector (usually not cost-effective indoors)

- Beam smoke detectors (in rare cases better than aspirating smoke detectors)
- Detector units or point detectors in the shade of ceiling beams left by daylight from windows
- Detectors coloured or painted in situ to blend with interior, decorations or surface
- Detectors detached from interior – demonstrating they are not part of heritage and reversible
- Different detectors combined to obtain the optimum of invasiveness and reliability

Reducing false alarms

These measures decrease the probability of false alarms while retaining response sensitivity to real fires:

- Linear heat instead of smoke detectors where applicable
- Aspirating smoke detectors instead of point smoke detectors
- Pattern recognition point smoke instead of standard point smoke detectors
- "2-of-any" or "double knock" detector logics for alarm or activation criteria
- Optimize smoke detection sensitivity setting (refer to G-JET or NFPA Appendix B information)
- Category and number of detectors (refer to G-JET or NFPA Appendix B information)
- Strategically locate sampling points/detector heads to avoid non-fire smoke, transmitters or other nuisance sources
- Prevention measures to control either the detector environment or to control malicious acts and human error are listed in the UK publication by ODPM, CACFOA and BFPSA[13] and are outside the scope of this report.

2.2 **Area fire detection** (outdoor)

An in depth study of this field was recently carried out by COWI on behalf of the Norwegian City of Fredrikstad and the RNDCH[6]. Following the evaluation project and full scale test series of ten area fire detection systems, and subsequent pilot installations of two line heat detector systems and one high sensitivity aspirating smoke system, it was concluded that all three systems were acceptable for use in the Old City of Fredrikstad. Therefore, the main criteria for selecting the best system became price. The aspirating system was clearly the least invasive, and the melting wire less invasive than the copper pipe system.

Tabled results may be used for guidance. But note that, in table A1 of Appendix A the criteria "Reliability and false alarms" will, if broken down in the Norwegian context, yield different results for wooden house areas of the country's interior with wood stoves versus houses in coastal areas. See Appendix A for time scale of events such as fire severity, alarms and a summary of the area fire detection results. The project was to provide knowledge and guidelines for similar wooden house areas in Norway.

3 INDOOR FIRE DETECTORS

3.1 Fire detection challenges in historical buildings

Fire detection installations in general are effective fire safety measures for heritage buildings and museums.

However, detector and cable installations present challenges:

- Irreversibly impair fabric or décor
- Renovation and maintenance incur irreversible damage to fabric or décor
- Aesthetically invasive in sensitive environments
- Detectors does not respond to fires as quickly as anticipated
- Detectors cause excessive nuisance alarms: get disconnected or disrespected
- Cable installations increase risk of fire from lightning
- Application may be inappropriate on terms of cost, efficiency, obtrusiveness

Chapter 3 and 4 provide known research results and recommendations on the above challenges, for indoor and outdoor applications respectively.

3.2 Detector categories

3.2.1 Traditional detectors (reference for rating less invasive detectors)

Point heat detectors
Point heat detectors have been in use for more than 100 years and are known for being robust, reliable and low cost - compared to smoke detectors which were introduced about 50 years ago.

Today's versions, however, contain communication, fault monitoring and some even pattern recognition printed circuits. Some have become prone to nuisance alarms, especially in harsh environments or from mobile phone or similar transmitter noise. But several products now seem to have greatly improved on this.

Generally, wireless ones are quite bulky, and those of smaller size are wired.

Point heat detectors appear underrated in performance as documented by NFPA and NIST, according to Bukowski's well known survey report on various international tests which have compared heat and smoke detectors with regards to life safety[19]. In the context of cultural value protection, however, we are looking for systems that are reliable and detect fire as early as possible to ensure fire damage is minimized, as long as they are cost-effective and minimum invasive. The Bukowski report challenged the conventional wisdom about smoke detectors being very much superior, and points out that heat detectors actually perform with adequate speed for many applications, even for life safety. The report supports findings of both laboratory and full scale tests (Vesterskaun test) of detectors performed by COWI (ICG) and SINTEF from 1984-1991.

However, point heat detectors do not present a good choice for non-invasive fire detection.

Point smoke detectors

Point smoke detectors are the workhorse of fire detection, very common and well known. In respect of this report, the appearance and installation of point smoke detectors is considered to be invasive in a heritage building, and our reference against which less invasive fire detection options are measured.

Detectors and wires are aesthetically intrusive by size, by number of units, by design or color. Fixings may cause irreversible damage to buildings or interiors.

3.2.2 Line heat detectors

These are small plastic tubes, small copper tubes, pairs of wires, optical fibres, melting wires or combinations. They are useful to detect fire in heritage buildings because they are installed unobtrusively and offer non-conducting types which remove the risk of fire being transmitted by services installations following lightning strikes.

They will generally raise an alarm well in advance of point heat detectors. Some are low cost and simple. A few have the advantage of integrating, i.e. low temperature along a line may cause an alarm as well as a tiny intense flame heating a spot on it only.

Figure 1: The stave churches are monitored by unobtrusive line heat detectors - and by aspirating smoke detectors.

The expensive ones may be sensitivity adjusted and address the fire location. The less expensive ones must be installed closer to potential fires, so require more length but less maintenance.

Contrary to popular belief the fixed threshold temperatures of line heat detectors are not very important: Lines are usually at the ceiling close to walls. Flames cling to walls and ceilings and have a temperature range of 500-900°C. Whether the threshold for activation is 68 or 150°C, or even 300°C is therefore less important. In fact, a melting wire for 180°C proved in a realistic test series[15] to respond as fast as 70°C point fire detectors and sprinkler bulbs - obviously because the RTI value of the wire is very low: The 2 mm diameter, non-insulated wire (it is to be applied non-insulated) is a low mass heat conductor that is heated throughout faster than most sensors. So, a most important performance parameter is time to heat sensors throughout (RTI value by NFPA, or Class ratings by EN 54 Standard).

The superior performance of line heat detectors versus point heat detectors is based on the fact that one detector line actually performs like thousands of point detectors next to each other. Other benefits of line heat detectors for heritage applications are the subtle way lines can be hidden, their robustness and their reliability.

Benefits to heritage provided by most line heat detectors

- Detection performance equivalent to one point detector at every point of line
- Very reliable
- Hardly visible at ceiling level
- May be run in ceiling voids with 2 mm diameter 50 mm drops only visible in rooms below
- May be run unobtrusively along cracks or shadowed areas
- May be fed into inaccessible voids via existing tiny holes or cracks
- Very robust
- Low cost
- Low cost maintenance

Working principles of line heat detectors

Water-filled plastic pipe at water mains pressure
Pipe rupture in fire causes water stream or pressure *drop* that triggers alarm or opens valves.

Air or gas filled plastic pipes at 2-4 bar pressure from canister or compressor
Pipe rupture in fire causes pressure *drop* that triggers alarm or opens valves.

Copper tube with pyrotechnical material
As pipe heats in fire material reacts to create a high pressure in the pipe which acts on a detection membrane element to trigger alarm/valves to open.

Stainless steel wire-size tube (2 mm diameter)
Internal 1 mm bore is air tight. By heating the trapped air *expands* to operate solid state circuitry to trigger alarm.

Copper tube (5 mm diameter)
Internal 4 mm bore is air tight. By heating the trapped air *expands* to operate membranes linked to switches.

Optical fibre
Several designs. Some are based on the signal transmission decreasing as heat from fire acts upon mechanical coils or supporting wires.

Melting wire
Very old principle. Typically a tin/lead alloy like solder wire is laid out in a loop with continuity monitored by electrical current through the loop. Circuit *breaks* when wire melts and triggers alarm.

Twisted pair of conductors
Monitored by electrical current. Each conductor insulated by material that decomposes at specific temperature, like 68°C. Circuit shorts immediately, as conductors are twisted with tension.

Cotton threads with weights, or with spring loaded switch
Used more than 100 years ago: Stretched along ceiling with tin boxes as weight. In fire thread breaks and boxes make noise as they drop to floor. Later type connects thread to spring loaded switch that triggers alarm by electric means.

Other designs
Some use bimetallic phenomena. Some use heat sensitive solid state materials. Some combine principles mentioned above.

Guide to select type of line heat detector

It is difficult to make precise recommendations as designs vary. Generally:

• Decide which fixed temperature thresholds or integrating sensitivity your application requires and go for the most robust, simple and inexpensive type.

• Consider if you require integrating types (no fixed temperature, slight heating over a long part of the line may trigger alarm as well as high temperature on a small part) or not. The integrating ones are expensive, may be overly sensitive for some applications and demand good maintenance.

• If the purpose is to detect flaming fire (smoke detectors are used for smoke) note that most flames exceed 500°C so threshold temperatures of 70 or 150°C may not make a big difference. Especially if lines are run at top of walls or under outdoor eaves.

• Consider if you need monitoring or not. Most are fault-monitored, but monitoring can induce faults by itself, and add cost and maintenance burden. If the design is reliable and may be tested routinely, opting not to monitor may be acceptable.

• Gas or liquid filled lines are pressurized and offer inherent fault monitoring, i.e. reduced pressure means either leakage, defect line or alarm - whereas the pyrotechnical, and locked atmospheric air types, need additional monitoring.

• Some line heat detectors appear quite invasive: They may exceed 5 mm in diameter, and they may involve bulky fixing points. The smaller ones are less than 2 mm diameter, require unobtrusive fixings and are hardly visible.

Figure 2: The historic Haverhill-Bath covered bridge in New Hampshire, built in 1829, was subject to arson fire in August 2002. A line heat detector system had been installed enabling the fire brigade to fight it quickly. Damage was minor (images by Protectowire).

3.2.3 Beam smoke detectors

Projected beam units consists of two components, a light transmitter and a receiver, that are mounted at some distance (up to 100m) apart. As smoke migrates between the two components, the transmitted light beam becomes obstructed and the receiver is no longer able to see the full beam intensity. This is interpreted as a smoke condition, and the alarm activation signal is transmitted to the fire alarm panel.

Figure 3: Beam smoke detectors may either cover approximately 100 m in direct line of sight, or cover similar distance by mirror(s) reflecting beam. Beam smoke transceiver sets for industry may appear obtrusive in heritage environments but can perform well if hidden (image by Autronica).

Beam smoke detectors are prone to false alarms or to become disabled due to misalignment, often caused by minor movements of the building construction or the mounting surface, but also due to infrared light saturation from sources of light. They frequently trigger false alarms when the beam is blocked by persons, draperies, ladders being carried or other objects. Since the advent of aspirating smoke detectors, conventional beam detectors tend not to be used nowadays, however in certain situations they can still offer a solution.

3.2.4 Aspirating smoke detectors

These are centrally mounted and draw air through pipes from protected areas.

There are two general types:
One type applies standard point smoke detector equipment in the fan box, the other applies very high sensitivity smoke detectors, nominal levels from 0.005 up to 20 %/m smoke obscuration. The latter is expensive but allows protection of large spaces, and is substantially more reliable.

Figure 4: Aspirating detection systems in churches and cathedrals typically involve pipes in attics and capillary drops through ceilings (image by Autronica).

Aspirating smoke detectors allow sensitivity settings over a broad spectrum, and the effective sensitivity may be varied from room to room within the same pipe and detector. Suction pipes may be hidden in voids, and tiny 5-6 mm "capillary" pipe branches may be fed through ceilings, hardly visible at floor level. Detection equipment can even be located in adjacent or remote buildings, connected with suction pipes underground, leaving nothing but a non-conductive thin pipe or hose in the protected building. This removes the risk of fire from lightning or high voltage arc.

Aspirating smoke detectors are the most robust smoke detectors available. They withstand physical impact, corrosion, wind, abrupt temperature variation, dust, spurious local sources of nuisance smoke, transmitters, electromagnetic noise and lightning arcs – while retaining optimum sensitivity to smoke from a fire.

Experience since 1993 has shown aspirating systems in a good light. They detect fires indoors when they are installed outdoors and vice versa, even if the building appears to be fairly tight. In a current pilot installation pipes of 20 mm diameter with suction holes along them will be installed outdoors and in attics with branching to strategic staircase tops.

The most promising application is to have a few hidden or freestanding outdoor suction holes only, such as in open air museums of small buildings. Aspirating types in general, offer the earliest response and best reliability.

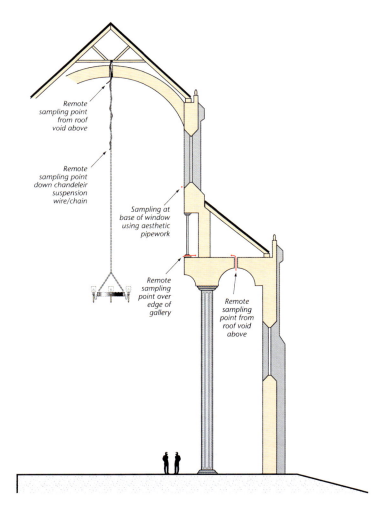

Remote sampling point from roof void above

Remote sampling point down chandeleir suspension wire/chain

Sampling at base of window using aesthetic pipework

Remote sampling point over edge of gallery

Remote sampling point from roof void above

Figure 5: Sample arrangement of points and pipework, running from aspirating smoke detectors in service rooms, for minimum invasion in cathedrals and large room volumes (drawing provided by AirSense Technology Ltd).

Figure 6: Suction point – barely visible (image by Elotec)

3.2.5 **Flame detectors** (detect thermal radiation)

These have become reliable for outdoor use, but are expensive - especially as a large number of units are required to "view" every spot where a flame may occur. In our automatic area fire detection (AAFD) test the flame detectors were directed towards windows where flames broke out and performed very well[6].

Figure 7: Flame detector (image by Hochiki)

3.2.6 **Visual image fire detectors** (detect visual light radiation)

Visual image fire detectors (VIFD) are ordinary surveillance cameras connected to software incorporating a fire detection program. The program analyzes the camera output signal for pattern recognition of fire signatures. These were developed in the early 1980s, and although the software has greatly improved, only 2-3 companies are known to provide such for fire detection.

Their inherent challenge is to filter out non-fire conditions that may appear like smoke or heat plumes in the software analysis.

They are typically used to detect smoke at an early stage within large room volumes or corridors, and in our context they are an option in galleries, museums and theatres. They are usually not the optimum choice either for mini-invasion or on cost grounds compared to aspirating detectors.

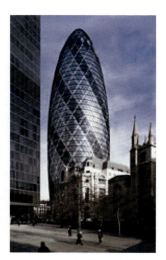

Figure 8: Swiss Re building, London, monitored minimum by VIFD (image by smoke D-Tec Ltd)

3.2.7 **Thermal image fire detectors** (detect thermal radiation)

Thermal image fire detectors (TIFD) are inherently much more reliable in detecting - compared to VIFD - as TIFD sense temperature, which is the only parameter the signal processing software needs to monitor in order to very reliably detect real fire in many applications. TIFD are more expensive, though, but most cost-effective in outdoor applications.

TIFD cameras are small, reliable and resistant for outdoor use, thus well suited for area fire detection. Cost is high, but has decreased consistently over recent years. Their economics are worthy of examination since they cover large areas, and do not need extra night lighting as visible light cameras do.

They may be combined with intrusion detection. The software analyzes the video-signal to detect both fire and intrusion by the same camera(s). If a minimum number of image pixels are in excess of for example 80 °C within a customer-defined part of the view area, an alarm may be raised automatically.

Until recently, it was not clear if TIFD's can be employed for effectively detecting hot smoke rising from a backyard or rear-side of a building, or if they require to "see the flames". The full scale test proved that the TIFD used would detect hot smoke (no flames in view) rising above the rear-side earlier than most line heat detectors and shortly after flame and aspirating smoke detectors. The camera was 80 m from the fire in this test, but can be installed closer or further away.

Eight brands of TIFD are known.

We currently consider TIFD's to offer considerable potential in AAFD. An installation with automatic detection software was recently commissioned at Røros World Heritage Site.

Figure 9: Claimed smallest advanced TIFD: Indoor version (held) and outdoor (fixed to surface) (photo by G Jensen)

NOTE:
See chapter 4 for further information about thermal image fire detectors, applicable to indoor and outdoor use.

3.3 Application Design Strategies

3.3.1 Minimum invasive solutions

Although most available detector categories are covered above, in most cases only the following types meet for minimum invasive detection criteria:

- Linear heat detectors
- Aspirating smoke detectors
- Wireless heat/smoke detectors
- Thermal* or visual cameras
- Beam smoke detectors**

*   *In special, usually large, indoor environments, thermal cameras are an option. But they are most efficient in outdoor applications – see chapter 4.*

**   *A good choice in 1980s, but since aspirating smoke detectors have generally proved superior.*

3.3.2 Wireless detectors

Wireless point detectors are an appreciated remedy to avoid unnecessary invasion. They have improved from earlier models which required frequent battery exchanges, except for non-heated areas in cold climates. The most expensive products offer high reliability and unobtrusive size, although most wireless units remain quite bulky compared with their hard wired counterparts.

Figure 10a: Sample wireless detectors: Point optical smoke (left), and aspirating sensitivity smoke detector (right) (images by Elotec).

Figure 10b: Early type wireless optical point smoke detector with antenna, in stave church (visible top right corner) (photo by G Jensen)

3.3.3 Selecting optimum smoke detector and sensitivity

Published guidelines for location of detectors tend to be generic to guarantee sufficient sensitivity and number of detectors in most applications and in worst case conditions.

This may neither lead to the optimum type of detector, nor to the optimum number of detectors or sensitivities being selected for a specific location. Sadly, most installations complying with guidelines turn out to be far from optimum in heritage buildings.

Updated standards, such as the NFPA 72, however, include the option to design fire detection systems from performance requirements, and provide guidance.

For smoke detectors, a free design program exists on the internet, G-JET[10] -at http://www.interconsult.com/G-Jet/ -see Appendix B. Using this tool one may select both the optimum smoke detector category and the optimum sensitivity at any given geometry and effective sensitivity. "Effective sensitivity" relates to how much smoke in grammes per cubic metres one will accept before an alarm is raised, at given applications. Default values are provided for typical room categories. G-JET applies formulae and accepted principles, such as stated in the appendix for engineering option in NFPA 72.

G-JET is very useful for detector design in heritage buildings by pointing out the optimum detector to be minimal invasive, adequately sensitive and practical.

3.3.4 Reducing false alarms

These measures decrease the probability of false alarms while retaining response sensitivity to real fires:

- Using linear or point heat instead of smoke detectors where applicable
- Using aspirating smoke detectors instead of point smoke detectors
- Using pattern recognition instead of standard point smoke detectors
- Optimize smoke detection sensitivity setting (G-JET or NFPA Appendix B)
- Refining design using engineeering option to determine category and number of detectors (G-JET or NFPA Appendix B)
- Strategically locating sampling points/detector heads to avoid non-fire smoke, transmitters or other nuisance sources
- Adopting preventative measures to control either the detector environment or to control malicious acts and human error such as those listed in the UK Government ODPM publication[13]. These are outside the scope of this report.

3.3.5 Smoke detector application designs

It is common that the wrong type of smoke detector is selected for applications. This stems from the industry standard of the 1970s when point smoke detectors were the only option, and all were made roughly equal in sensitivity. Later, as detection technology became more sophisticated, beam smoke and aspirating smoke detectors emerged, and widely adjustable point smoke detector sensitivity was made possible. This was not necessarily matched by new thinking from the electronics engineers handling alarm systems who would have required more specialized knowledge to understand smoke buoyancy, characteristics, development and spread – so as to be able to capitalize on equipment developments.

Figure 11: The Norwegian Museum of Cultural History. Typical surroundings for smoke detector application design engineering.

Today, design guidelines exist to handle these variables and this offers benefits to heritage. Given the required sensitivity for an application:

- The least invasive smoke detector category is selected
- The least invasive smoke detector location is selected
- False alarms are kept to a minimum
- Both smouldering and flaming fires will elicit responses as early as expected

Simple formulae for smoke dilution are provided by NFPA Standard 72[9] and the SFPE Fire Safety Engineering Handbook[25], but no European standard is known to include this yet. The internet available program G-JET[10], previously referred to in section 3.3.3, can also be used to calculate and suggest appropriate category of detectors and their initial sensitivity settings for indoor applications.

4   OUTDOOR FIRE DETECTORS

4.1   Area fire detection

The term "automatic area fire detection" (AAFD) means a system to detect fires within a courtyard or group of buildings which are at risk from external fire spread, without necessarily locating any detector or equipment inside the buildings. The aim is cost-efficiency and to reduce physical and aesthetic invasion.

Several studies and tests have been made in Norway to explore the possibilities. The performance requirement is to detect fire as early as possible to ensure the fire brigade manage to prevent conflagration. It should be noted that with such a set up the loss of small single houses might be accepted when designing to protect groups of buildings from conflagration.

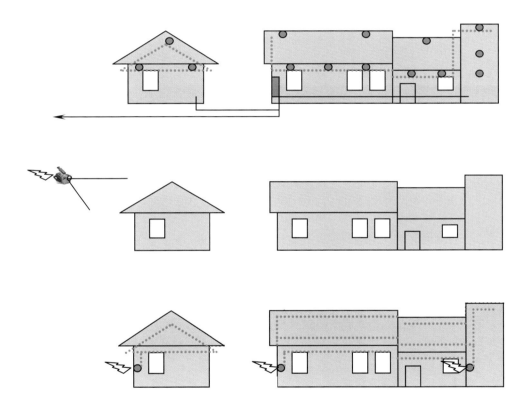

Figure 12: The sketches illustrate that traditional wired point detectors require significant invasive installation of cables and units(top), whereas thermal imaging camera does not at all (middle). Combination wireless units and line heat detection is a reasonable solution (bottom).

4.2 Detector Categories

4.2.1 Thermal image fire detectors

Thermal image fire detectors (TIFD) are cameras that sense heat radiation rather than visual light, and include software that automatically detects and raises the alarm if a number of pixels exceed set thresholds. For example "no more than 125°C anywhere" or "no more than 12 pixels exceeding 70°C within this region". A number of regions can be defined within the live picture display, as rectangles or circles etc, and specific alarm response criteria are assigned to each.

Figure 13: TIFD camera for outdoor use
(photo G Jensen)

TIFDs are expensive but superior to visual image fire detectors because they:

• Are not affected by shadows and light, moving objects or persons*
• Display the same picture day and night
• Respond to heat radiation only which is the only parameter to be monitored
• Thresholds are set in temperature only, so allow simple software to be used

> * *TIFD may be set up for detecting intrusion by persons or vehicles etc at the same time or exclusively, and this is the most common application of 24hr thermal imaging camera monitoring*

TIFDs are reliable in detecting outdoor fire in varying weather conditions and in night as well as day time. Live images can be transmitted to 24hr manned stations. Some include visual cameras, and one product fully integrates both cameras to enable a seamless blend of both visual and heat images easily adjusted by operator.

Figure 14: Thermal imaging camera for evaluation outdoors
(photo G Gjelseth)

TIFDs are useful for heritage applications because:

• One single camera may monitor one ancient city centre or group of houses (although two pan/tilt cameras, or 2-4 fixed cameras, is recommended).

• Fires which may escalate into conflagration are detected before it is too late for the fire brigade to control them. The system can even identify by heat radiation of smoke plumes extending above roofs, escaping through broken windows from internal fires which are out of camera view.

- No installation at or inside any protected building is required.

- Operators verify the potential fire by assessing the live pictures from TIFD before responding. Thus, high sensitivity is allowed without causing unnecessary fire brigade responses.

- TIFDs have internet (IP)adresses to allow real-time pictures to be viewed by rescue services, museum staff or key personnel with password identification from vehicles or at home so rescue efforts are optimized during fire.

The drawback of TIFDs is cost - they are very expensive if used for a single object. For groups of houses, though, TFID is easily less expensive than detectors in every house.

As with any image fire detector, one has to assess the risk and if necessary install additional detectors indoors, or on facades out of view of cameras where indoor fires or fires at facades are critical in respect of conflagration, should they not be detected early enough by camera alone.

Time of intervention, fire brigade capacity, size of buildings and above all the inter-building fire spread properties are parameters of analyses to be made before installing TIFD's. Groups of relatively small houses and possibilities to locate cameras high above lend themselves for effective TIFD applications.

Figure 15a: Images by TIFD (website advertisements, various manufacturers)

Figure 15b: Compact TIFD+VIFD for outdoor use (image by SRA/Mikron) Detector is a combination TIFD+VIFD camera for 24hr outdoor monitoring. It turns 360 degrees, pans and tilts. Upon alarm the operator may evaluate the conditions by thermal or visual image, or by any seamless combination of the two. The camera moves continuously to preset positions and detects any exceeded temperature limits. 255 positions are possible. Each position allows 32 zones defined with their own criteria. These are installed to monitor wooden towns in Norway.

Figure 16: The unique fishing community heritage site at Grip Island in Norway is one application for TIFD. Planned camera viewing angles shown.

4.2.2 Aspirating smoke detectors

See chapter 3 for a description of aspirating smoke detectors in general, and chapter 5 for application test results. These detectors are amazingly effective for reliable area fire detection and second to thermal image fire detection only in terms of offering minimal invasive solutions.

Any serious initial fire that may threaten to grow into a conflagration produces large quantities of smoke, some of which tends to migrate between buildings and into the buildings. As the principle of aspirating smoke detectors is to sense smoke from all its suction points, they are very sensitive if all points provide some smoke. Even very low concentrations of smoke in all or many of the points add up to a substantial reading at the sensor which triggers the alarm.

In contrast to point smoke detection, aspirating detectors detect smoke even if it drifts away from the origin of the fire and becomes diluted.

Sampling pipes with suction holes may be installed in attics and in the top of staircases only, offering out of sight reversible installations. In these locations they can detect smoke from fires within the building, externally or in the neighbourhood. This principle has been tested and proved very effective.

Aspirating smoke detectors are not recommended for districts where large amounts of smoke are normal, such as in the UNESCO World Heritage site of Røros with wood stoves very much in use. For most other areas, however, the sensitivity can be adjusted to perform satisfactorily.

Figure 17: Aspirating smoke detector pipe hidden in a tree at fullscale test [6] (photo G Jensen)

4.2.3 **Line heat detectors**

See chapter 3 for a description of line heat detectors in general, and chapter 5 for application test results. Line heat detectors have proven reliable for outdoor or area fire detection, except for some early developments and types that can be easily damaged when installed.

The thinnest types are the least invasive, while the thicker ones and their fixings may be quite obtrusive.

In excess of 10 types have been evaluated. One test[6] included a pneumatic type, a paired wire and a melting wire type for subsequent pilot installation in the old city of Fredrikstad. The results were promising, and all three were found acceptable for area fire detection.

The first line heat detector applied at the Røros world heritage site, a melting wire type, proved sensitive to the way it was installed and unfortunately tedious repair of the initial installation was required. An aesthetically more invasive, but more robust type (insulated pairs) has been installed to cover the remaining parts of the village.

Line heat detectors for area fire detection at wooden town centres are typically installed outdoors under eaves and indoors in attics and other critical rooms in order to avoid inter-building conflagration.

Figure 18: Sample images of area detection at historical site (mining town, Røros). Solid 2 mm melting wire line heat detection hardly visible, indoors and outdoors. Large buildings get supplementary indoor optical point smoke detectors. All wireless line and point detectors transmit alarms to fire station (photos, G Jensen)

4.2.4 Flame detectors

See chapter 5 for application test results. Flame detectors are relatively well known and proven for outdoor or indoor applications. For single object monitoring flame detectors are reliable, but in a townscape context they are less useful because they are expensive and less reliable. This is because the sensing of flames requires direct sight and it is difficult to provide full cove rage with out a large number of detectors.

Figure 19: Flame detector in full scale testing[6]

4.2.5 **Point smoke detectors**

See chapter 3 for a description of point smoke detectors in general, and chapter 5 for application test results.

Optical point smoke detectors with sensitivity at 5-15 %/m obscuration depending on geometries and quiescent smoke levels could theoretically be applied for area fire detection if located at strategic locations such as under eaves, in attics, stair-cases, overhangs and so on.

However, point smoke detectors do not cope well with outdoor climates and they are not readily available with these sensitivity settings (sensitivity of standard point smoke detectors are typically within 1-3 %/m obscuration), so aspirating types are much preferred.

Point smoke detectors are effective in complementing outdoor area fire detectors when smoke detectors indoors are required for large buildings in the area, or in buildings in which fires very quickly may grow beyond the control of fire brigades.

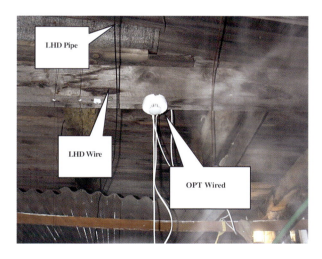

Figure 20: Wired point smoke detector beside two line heat detectors during full scale test[6] (photos G Jensen)

4.3 Application Design Strategy

Parameters to consider in area fire detection design are listed in Appendix A.

The alarm must be transmitted to the fire brigade as early as possible to prevent a fire from escalating to conflagration. This is very different from the usual task of protecting lives or valuable building contents or protecting a single building. The time to intervention is of paramount importance, as is the fire brigade capacity. Furthermore, the expected inter-building fire spread velocity is decisive.

Even if all involved buildings are sprinklered, this is not an effective protection from conflagration if the problem is easily ignited outer claddings. Once a fire is established outdoors it may spread to other buildings and even if sprinklers inside operate they may not stop the fire for two reasons: sprinklers are not designed to stop intrusive fires and if mains fed sprinkler heads are activated in several buildings at once the water supply quickly becomes inadequate.

It may be tempting to select the most sensitive detector, or put as many as possible indoors. But this may be very expensive and give rise to nuisance triggering of false alarms, while still missing the performance requirement of alarming as early as required to prevent conflagration, which in area fires relates to the fire conditions externally.

When dealing with heritage buildings, account should be taken of progress in developing more simple detectors which lead to less invasive installations. Appendix A offers guidance, but check for new, improved products. Consideration must also be given to how often the installation needs to be renewed, or how often it needs attention which involves ladders etc. Over time frequent upgrading and maintenance may be damaging to the subject being protected. If so, one should rather avoid the "best" detector in order to await new technology, or select the second best overall if it is less invasive.

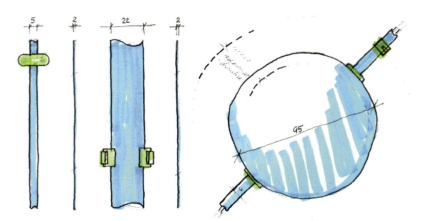

Simple comparison of invasive impact: relative

Figure 21: Size of area fire detectors. Left to right: Line Heat Copper Tube (5 mm diameter + clips), Line Heat Stainless Steel (2 mm diameter), Aspirating Sampling Pipe (22 mm diameter + clips), Solid Melting Wire (2 mm diameter) and Point Smoke Detector (95 mm diameter with 6 mm diameter cable + clips) with Cable (6) (ill. by COWI)

5   EVALUATION TESTS

5.1   Drammen test series

In 1993 a test series[22] was performed to determine properties of available smoke detectors for applications in the stave churches and other churches. The tests were made in a barn, using realistic fuels and both flame and smouldering combustion. In addition to common point detectors, aspirating smoke detectors were included, but no beam smoke detectors. At around the same time outdoor detectors were tested at another facility: See section 5.8.

The results were later confirmed by other test series described in this chapter 5. The Drammen tests initiated the G-JET model[10] development (section 3.3.2).

5.2   Vennesla test series

The most extensive realistic full scale test of both area and indoor fire detection was made at the demolition burning of a large wooden building in Vennesla, Norway[15].

A melting wire fire detector was evaluated against a series of other detectors.

Figure 22: Test building at Vennesla

Proprietary detectors and sprinkler equipment used in test:

Home Smoke Alarms:
RI_D = Ionization smoke (manufacturer D)   (reference detector)
RI_F = Ionization smoke (manufacturer F)   (reference detector)

Detectors Wired to Alarm Panel:
SFW = Line heat detector, (meltable low mass wire type, 183 °C)   (test subject)

V =   Point heat detector (Class 1 EN 54:5)   (reference detector)
I =   Ionization point smoke detector (EN 54:7 and 9)   (reference detector)
O =   Optical point smoke detector (EN 54:7 and 9)   (reference detector)

Sprinklers:
Residential 1: Residential Pendent (ceiling) *(57-77 °C – RTI <50)*   (reference "detector")
Residential 2: Residential Concealed (flush ceiling) *(57-77 °C – RTI <50)*   (reference "detector")
Residential 3: Residential Sidewall (sidewall) *(57-77 °C – RTI <50)*   (reference "detector")
Standard spray ceiling (red bulb) *(57-77 °C – RTI >80)*   (reference "detector")
Standard spray ceiling (yellow bulb) *(79-107 °C – RTI >80)*   (reference "detector")

Fire test scenarios[15]:

Fire A	*Attic*	*Optional, not performed*
Fire A	*Attic*	*Optional, not performed*
Fire B	*Clothes Dryer*	*Clothes within were ignited, door left open.*
Fire C	*Bookshelf*	*Not performed (refer to D1 and D2 tests below)*
Fire D1	*Padded sofas etc*	*This test involved all types of detectors and sprinklers in the same living room.*
Fire D2	*Padded sofas etc*	*As D1, but later stage intervention to extinguish.*
Fire E	*Arson fire outdoor*	*Two dry standard pallets sloping upwards against wooden external façade. Saw dust in a 50/50 % paraffin/diesel mixture used to ignite.*
Fire F	*DJ-stage in assembly hall*	*Clothes, draperies, padded chairs, power amplifiers and more.*
Fire G	*Entrance, outdoor*	*Children playing with matches-type of fire: Plastic, wooden sticks, milk boxes (very small fire).*
Fire H	*Demolition fire*	*Those detectors which survived all previous tests monitored during demolition fire - compare D2.*

Conclusions: If installed underneath outdoor eaves and overhangs in courtyards - with loops inside of attics, at tops of staircases and inside of rooms with no windows - in wooden town centres like the Old City of Fredrikstad, Norway, melting wire line heat detectors will:

- Respond to all flaming fires outdoors and indoors sufficiently fast to prevent conflagration
- Respond before, or at the latest at time of indoor flashover
- Respond within 10 seconds following arson fire ignited with liquid outdoors
- Respond within 50 seconds to slow fire outdoors (professional guesstimate: at heat release rate of <100 kW if line detector <2 m above source)

Figure 23: Fire D1 just after Ignition

Figure 24: Fire F just after ignition

From assessment by the live event, recorded video and images it was judged that at the time of line heat detector response, fires would be easy to extinguish, and that intervention times of 5-10 minutes can be allowed for whilst still ensuring conflagration is avoided.

It was also concluded that smoke detectors hardly make a substantial difference: The worst case scenarios involve arson fire with liquid and these develop so fast that line heat versus smoke detector responses would differ by typically 5-10s only. Though smoke detectors respond to smouldering fires and may trigger an alarm perhaps hours ahead of heat detectors, such fires are not critical in terms of conflagration at that stage, and are contained within rooms so even after flashover they are fairly easy to suppress[15].

5.3 Svorkmo test series

See Appendix A for a detailed description and event diagram of this test series [1,2,6] which involved area fire detection based on indoor as well as outdoor detectors. Summaries of all area fire detectors tested and their relative responses is shown in the tables and figure of Appendix A.

The objective of the test series was to optimize minimum invasive solutions, reliability, cost and practical concerns like avoiding access into private homes.

5.4 Haugesund test series

A student exercise report to test and evaluate line heat detectors for indoor and outdoor use in historical buildings[16].

The report generally confirms other data, but is very useful also in terms of its realistic and careful considerations of practical implications.

Figure 25: Image upper right shows 4 line heat detectors in test, from uppermost: Stainless steel tube (2 mm), Copper pipe (5 mm), Meltable wire (1.6 mm) and detection/extinguishing fluid filled plastic pipe (8 mm). Shortcircuiting pair with melting insulation type (4 mm) were not involved in full scale test[16].

5.5 NIST home smoke alarm tests. Vesterskaun test series

An on-site realistic test of home smoke alarms was performed in 1989[23].

A comprehensive laboratory test series was performed by NIST (US) [23] and completed in 2004.

These projects represent valuable knowledge on the performance of point fire detectors, though not on such properties as minimum invasiveness. Both test projects involved heat, optical and ionization smoke point detectors.

5.6 Røros and Rødven tests of outdoor thermal imaging detection

A project for the Norwegian Directorate for Cultural Heritage and the Directorate for Civil Protection and Emergency Planning, involved pilot installations of thermal image fire (camera) detectors which were monitored for several weeks in rough climates, one at a small church and one in a church tower overlooking the ancient mining town of Røros.

The report[7] concluded that thermal image fire detection fulfilled all expectations of performance. Practical problems with transmitting and recording the signal are no longer valid in the compact solutions offered today by manufacturers. Thermal image fire (camera) detection was found to be superior to both flame detectors and visual image fire detectors in most respects, but more expensive. Following this project a set of performance requirements for thermal image fire detector systems was developed, and since refined in the course of subsequent commercial competitive tender exercises.

5.7 Melhus ad hoc test

During fire demolition of a domestic wooden house evaluation of smoke alarms and line heat detectors were made, using realistic tests.

The detectors were installed as planned for similar houses of the UNESCO world heritage site of Røros.

Figure 26: Line Heat Detectors at ceiling for test

It was concluded that the faster the fire development, the less important the location of line heat detectors in the vicinity[11]. When line heat detectors are fitted to all outdoor eaves, and passed through into most rooms at the ceiling corners of window walls, the overall response to fires indoors compared well to that of typical smoke alarms in corridors and staircases only. Furthermore, response times would be substantially reduced compared with installations involving outdoor and attic coverage only, as deducted from the Vennesla test.

It was confirmed that non-insulated melting wires of 2 mm diameter and melting point of 183°C respond as fast as common fixed temperature line heat detectors of nominally 70°C. This is due to the very low RTI-value (not measured) of the melting wire design.

5.8 Application tests for stave churches

Electric line heat detectors
Various types of electric line heat detector types have been in use since 1982 at the stave churches. The best reliability is currently obtained from a system using short circuiting paired wires with water tight outer shield.

Pneumatic line heat detectors
Pneumatic line heat detectors were tested at 1:1 scale mock up stave church tests in 1993. A 5-10 mm copper tube with pyrotechnical material inside was among those tested. When heated the material "ignites" and the pressure build up automatically operate sensors for alarm, and a deluge or preaction valve for extinguishing systems. It worked satisfactorily but was eventually abandoned in favour of pressurized plastic pipes activating likewise and monitored. For the latter, both a standard proprietary type and customized systems were applied.

The plastic pipes were smaller and easier to fit unobtrusively, and did not need renewal every 5 or 10 years as with the copper tube. But, the plastic material proved critical as most types leak gas. Also, connections are prone to leak and both leak modes may cause excessive load on compressed gas containers or air compressors that compensate for leaks. The low monitoring pressure makes the valve activation very sensitive, causing nuisance activations. In at least one case this caused activations to partly fail on real fire demand.

It should however be noted that the experience outlined above is from outdoor use in harsh environments with temperatures typically lower than minus 30°C.

Aspirating smoke detectors
Part of the Drammen test series reported above included proprietary aspirating smoke detectors. These were some of the very first tests world-wide in large room volumes with high ceilings employing this detection principle. Several important benefits were demonstrated (compared to point smoke detectors):

- Easy to fit unobtrusively/minimum invasive (see sections on aspirating type)
- Very high sensitivity
- Very low sensitivity to false alarms
- No electrical conductors above ground level so reducing risk from lightning
- Detect fires both indoors and outdoors efficiently

As the pilot installations were made, tests on site were performed, and routine smoke alarm tests developed specifically for the stave churches. The mock up and in situ tests proved useful for system designs:

- Detection points at several heights to compensate stratification (smoke layering) and insufficient buoyancy to reach ceiling mounted detectors.
- Outdoor façade detectors became obsolete as the aspirating detectors inside worked better – this relieved the stave churches of substantial wire invasion
- The tests initiated development of the G-JET tool (See Section 3.3.3. and Appendix B).

Drawbacks - although they are improved much since initial introduction - noise nuisance in churches caused by detector fans was found to be a problem. Routine exchanges of dust filters may also be necessary quite frequently, and installations must be made carefully to avoid condensation problems in non- or part-heated buildings like the stave churches.

Aspirating smoke detectors from a number of manufacturers have performed impressively well overall in the stave churches.

The information set out in this section, 5.8, is drawn from installation projects, ad hoc tests at the sites and reported in various documents linked to the installation projects performed by COWI on behalf of the RNDCH.

Figure 27: Fire detectors hardly visible at Norwegian Folk Museum, Oslo. Line heat detectors applied in shades of outdoor eaves and in shades behind ceiling beams indoor. Point smoke detectors painted to blend with surroundings in shaded areas both indoor and outdoor.

6   FUTURE DETECTION CONCEPTS

In the long term perspective of historic buildings it is pertinent to consider current technology and which advancements may come. Invasions into building fabrics today may appear wasted in the future if less invasive detector concepts become reality. Below, sample concepts now being developed are described.

6.1   Structural borne sound fire detectors

Acoustic sensors fixed to structural members of a building detect flaming fires by way of specific algorithms. This is explored and reported[26], but products are not yet known to be available.

Piezoelectric transducers can scan a large area and are unaffected by the presence of people or machinery, both of which can sometimes give false clues of fire. The sounds of thermal expansion spread more quickly than combustion products or infrared radiation (heat). An acoustic sensor may serve in an integrated, intelligent detection system, which can locate and analyze hidden hot spots in a building[27].

A potential benefit is that detectors may be mounted outside of sensitive rooms, where they are hidden from view and do not involve irreparable or aesthetical damage by fixings etc. Such detectors are probably best used in combination with other sensors for multi-criteria evaluation.

6.2   Airborne sound fire detectors

Microphones detect fires by characteristic airborne sound from the fire source. Fire pattern recognition microphone detectors are not known to be marketed for this purpose, but they are interesting components of multi-criteria concepts. Microphone recordings were made at the Svorkmo tests, Jensen[6].

For example, microphones in remotely monitored smoke detectors may help monitoring station personnel to verify if there is cause to intervene or not. Likewise, small web cameras in monitored rooms may incorporate microphones.

Due to the small size and working principle, potential benefits are that detectors may be less intrusive, mounted out of view or incorporated into other units without changing the appearance of the units.

6.3 Advanced multi-criteria fire detection

Signatures of initial fires are not only heat and smoke particles. Signatures are also a huge range of various gases, smells, air- or structural borne sound, electromagnetic radiation (including visible light) and much more.

Along with very small and inexpensive new semiconductor sensors developed lately, detector manufacturers are working on algorithms in detector software for pattern recognition of real fires and to distinguish them from nuisance conditions.

Multi-sensor, multi-criteria, fuzzy logic and algorithm pattern recognition are terms describing detectors on the market already. These, however, are just starting to explore the possibilities. If one combines sensors for gases, artificial nose sensors, humidity sensors, temperature vs time curve sensors and so on - the resulting algorithms potentially become very complex, but also potentially extremely reliable in evaluating weather-conditions warrant sounding a fire alarm or not.

The tiny sensors involved may compensate for sensors that otherwise are invasive, and reliability may increase. Therefore, future pattern recognition concepts should be of interest for heritage applications.

REFERENCES AND BIBLIOGRAPHY

1   Jensen, Geir. Reitan, Arvid. Bøifot, Martin: *Gamlebyen Fredrikstad Brannsikkerhet. Områdebranndeteksjon: Alternativene. (Area fire Detection Alternatives for the Old City of Fredrikstad).* Komplett versjon 31.01. 2000. InterConsult Group. 2000. (Norwegian language).

2   Jensen, G. Reitan A. M. Bøifot: *Fire Protection of Old Fortress Town Fredrikstad: Area Fire Detection – Full Scale Tests and Evaluation of Systems.* Norwegian language (Svorkmo). ICG. 1999.

3   Jensen G, Tørlen Lønvik E, Heskestad A W: *Application Specific Sensitivity – A Simple Engineering Model to Predict Response of Smoke Detectors.* AUBE '99 Conference, Duisburg, Tyskland. 1999.

4   Tørlen Lønvik, Jensen, G.: *Fire Detection and Automatic Operation of Extinguishing Systems – designed for a Harsh Climate and a Tender Environment.* ICMS Conference, Stavanger. 1995.

5   Jensen, G., Landrø, H.: *Novel Techniques for Active Fire Protection of Historic Towns and Buildings.* International Conference on Fire Protection of Cultural Heritage.Thessaloniki. 2000.

6   Jensen, G.: *Gamlebyen Fredrikstad Brannsikring sluttrapport: Konklusjoner etter utredning, tester og 9 mnd pilotprøving. (Fire Protection Summary Report following Surveys, Tests, 9 month trials at the Old City of Fredrikstad).* Interconsult. 2001. (Norwegian language).

7   Jensen, G.: *Varmekamera for deteksjon av brann og ferdsel (Thermal Imaging Detection of Fires and Intruders).* 2002. (Norwegian language).

8   Jensen, G.: *SLUTTRAPPORT: Tester i Røros, Rødven, Trondheim (Final Report on Thermal Imaging Detection following Tests at Rødven, Røros and Trondheim).* Interconsult ASA. 2003. (Norwegian language). http://www.dsb.no

9   National Fire Protection Association: *NFPA 72: National Fire Alarm Code.* National Fire Protection Association, Quincy, Massachusetts. 1999

10 Interconsult ASA: *G-Jet: A Free Smoke Detection Application Calculator.* Interconsult ASA, Trondheim. (Norwegian, German, English languages). http://www.interconsult.com/G-Jet/. 2003.

11 Steen-Hansen, Jensen, Hansen, Wighus, Steiro, Larsen: *Byen brenner! (City's on Fire!).* SINTEF Report A03197 on behalf of The Directorate for Civil Protection and Emergency Planning, Norway. 2004. (Norwegian language).

12 Artim, Nick: *An Introduction to Fire Detection, Alarm, and Automatic Fire Sprinklers.* Northeast Document Conservation. Technical Leaflet.

13 *A guide to reducing the number of false alarms from fire-detection and fire-alarm systems.* CACFOA.BFPSA and Office of the Deputy Prime Minister. (London, 2004).

14 Steen-Hansen, Jensen, Wighus, Steiro: *Fire Protection of Røros: A Historic Town on the UNESCO World Heritage List.* Interflam 2004.

15 Jensen, G.: *Full-Scale Test Series of A Novel Line Heat Detector. Interconsult on behalf of Spider Fire Wire.* 2001. (Norwegian language).

16 Sandvik, Sandal, Grimstvedt: *Evaluation of Line Heat Detectors (Based on Full Scale Tests and Literature).*Stord/Haugesund University College. 2002 (Norwegian language).

17 Jensen, G.: *Brannsikring av fredede bygninger og historiske trehusmiljøer: Hvilke tiltak finnes, hvilke er best egnet, og hvilke erfaringer har vi? (Fire Protection of Listed Buildings and Wooden town Centres: What Measures are Available, which is Best and what is theExperience?* Paper/report. Norsk Brannvern Forening. Interconsult. 1998. (Norwegian language).

18 *Nytt system for deteksjon (A new Fire Detection System)* (Firesys – pilot-prosjekt Norsk folkemuseum). Sikkerhetsmagasinet. 2/01. (Norwegian language).

19 Bukowski, R.W.: *Studies Assess Performance of Residential Detectors.* NFPA Journal Jan/Feb 1993.

20 Patton, Richard M.: *A rebuttal to Bukowski's article on detectors.* NFPA Journal Jan/Feb 1993.

21 Jensen, Lønvik, Landrø: Making *the Right Decisions on Smoke Detectors.* Safety4Industry.com. Interconsult. 2001.

22 Jensen, G.: *Deteksjon av røyk i kirker (Smoke Detection in Churches).* IGP AS (COWI). 1993. (Norwegian language).

23 Meland, Lønvik: *Detection of Smoke.* SINTEF STF 25 A89010. Norway 1989.

24 Bukowski, Richard W: *Home Smoke Alarm Tests.* Fire Research Division. National Institute of Technology, USA. 2004.

25 *The SFPE Handbook of Fire Protection Engineering.* Third Edition. 2002. The National Fire Protection Association, and Society of Fire Protection Engineers.

26 Grosshandler, Jackson: *Acoustic Emission of Structural Materials Exposed to Open Flames.* NISTIR 4984. Building and Fire Research Laboratory. National Institute of Standards and Technology Gaithersburg, Maryland, USA. 1992.

27 *Listening for hidden fires – acoustic sensors can detect sound of materials about to burst into flame.* Science News. July 1993.

28 Massingberd-Mundy, Peter: *Sampling Points (Outline use of aspirating smoke detectors for heritage buildings).* Fire Engineering Journal. February 2004.

Summary of area fire detection

PERFORMANCE OF AREA FIRE DETECTION AND INTERVENTION

From literature and application test experience it is concluded that very simple fire detectors may offer acceptable safety against conflagration. Expensive detectors tend to be costly to maintain and more susceptible to false alarms and downtime due to failure or maintenance. This may impact on response to real alarms. Furthermore, the expensive solutions - in general - require more operator skills, training and/or involve more aesthetic and physical invasions. If fire signatures can be detected from the outside by, for example, an abnormally hot window pane, or smoke migrating from the building (thermal image fire detection is able to do both), and time to fire brigade intervention is short, acceptable protection against fire escalating into a conflagration may be provided without resorting to extensive indoor protection.

Figure A1 shows response time points for both internal and external detectors at a wooden house during demolition fire[6]. The type of detectors, their locations and response times are further listed in the comparison table below. The table is intended as a guide only.

Table A1: Ref[6] provides details for optimum use of this table. Detectors were not positioned for real life full coverage - they were positioned within a limited area opposite the house corner where the fire started. Fire brigade initiated fire fighting 30 minutes after ignition and gained control after less than 5 minutes.

Rating 1-5: 5 most favourable 1 least favourable	Point smoke detector, optical	Aspirating smoke detector	Aspirating smoke detector, high sensitivity	Line Heat detector, pneumatic type *Type I Intergrating*	Line Heat detector, pneumatic type *Type II Intergrating*	Line Heat detector, melting wire type *Fixed Temperature*	Line Heat detector, short-circuiting type *Fixed Temperature*	Flame detector	Thermal imaging fire detector	Compact wireless all-in-one detector
Detection performance	4	4	4	4	4	4	4	3	4	4
Reliability, false alarms	2-4	3	4	4	4	3	4	3	-	2-4
Integrating	2	4	4	4	4	2	2	2	4	2
Adjustable Sensitivity	2	4	4	4	3	2	2	4	4	2
Physic/Aesth Invasion	3	3	4	3	4	3	3	4	5	3
Locating fire point	5	3	2	3	3	4	3	4	4	4
Access into private homes	4	2	3	2	2	4	3	3	5	4
Reversibility	3	3	4	2	3	5	3	4	5	3
Routine Realistic Tests	4	4	4	3	3	4	3	3	3	4
Flexibility	2	3	3	2	2	4	3	1	4	3
Accessibility Spare Parts	4	4	4	3	2	4	4	2	3	3
Skills to Maintain	3	3	3	2	2	4	4	3	2	4
Ownership	4	2	3	2	2	4	3	2	5	4
Operational Life Time	3	3	3	4	4	2	3	3	3	3
Cost Investment	3	3	3	3	2	4	3	2	4	4
Cost Operational	3	3	3	3	3	4	3	3	3	3

Test installations and evaluation of thermal image fire detectors were performed later, see ref notes[6].

	>20 min	>15min	>10min
Point smoke detector indoor, optical (2. level)	•		
Line heat detector, pneumatic type, indoor (prewarn)	•		
Aspirating smoke detector, indoor	•		
Aspirating smoke detector, high sensitivity, outdoor	•		
Point smoke detector, optical, indoor (3. level)		•	
Flame detector, outdoor		•	
Thermal imaging fire detector, outdoor (80 m distance)		•	
Line heat detector, melting wire type, outdoor		•	
Line heat detector, pair of wire, outdoor			•

Table A2: Time margins provided by various area fire detection systems - from test during fire demolition of a 3 storey wooden house (Svorkmo, see refs note[2])

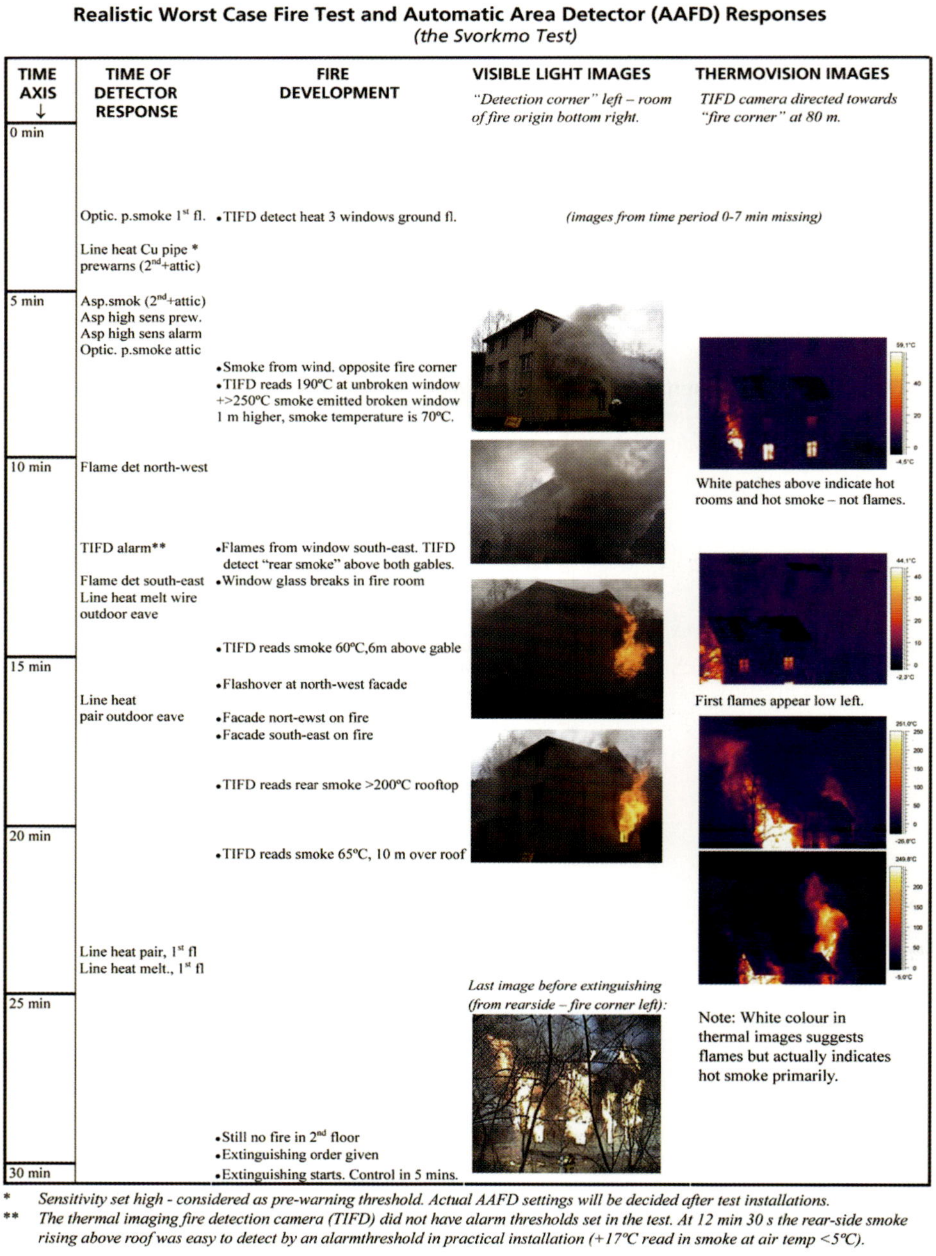

Realistic Worst Case Fire Test and Automatic Area Detector (AAFD) Responses
(the Svorkmo Test)

TIME AXIS ↓	TIME OF DETECTOR RESPONSE	FIRE DEVELOPMENT	VISIBLE LIGHT IMAGES *"Detection corner" left – room of fire origin bottom right.*	THERMOVISION IMAGES *TIFD camera directed towards "fire corner" at 80 m.*
0 min	Optic. p.smoke 1st fl.	• TIFD detect heat 3 windows ground fl.	*(images from time period 0-7 min missing)*	
	Line heat Cu pipe * prewarns (2nd+attic)			
5 min	Asp.smok (2nd+attic) Asp high sens prew. Asp high sens alarm Optic. p.smoke attic	• Smoke from wind. opposite fire corner • TIFD reads 190°C at unbroken window +>250°C smoke emitted broken window 1 m higher, smoke temperature is 70°C.		
10 min	Flame det north-west			White patches above indicate hot rooms and hot smoke – not flames.
	TIFD alarm** Flame det south-east Line heat melt wire outdoor eave	• Flames from window south-east. TIFD detect "rear smoke" above both gables. • Window glass breaks in fire room • TIFD reads smoke 60°C,6m above gable		
15 min	Line heat pair outdoor eave	• Flashover at north-west facade • Facade nort-ewst on fire • Facade south-east on fire • TIFD reads rear smoke >200°C rooftop		First flames appear low left.
20 min		• TIFD reads smoke 65°C, 10 m over roof		
	Line heat pair, 1st fl Line heat melt., 1st fl		*Last image before extinguishing (from rearside – fire corner left):*	
25 min				Note: White colour in thermal images suggests flames but actually indicates hot smoke primarily.
30 min		• Still no fire in 2nd floor • Extinguishing order given • Extinguishing starts. Control in 5 mins.		

*   *Sensitivity set high - considered as pre-warning threshold. Actual AAFD settings will be decided after test installations.*

**   *The thermal imaging fire detection camera (TIFD) did not have alarm thresholds set in the test. At 12 min 30 s the rear-side smoke rising above roof was easy to detect by an alarmthreshold in practical installation (+17°C read in smoke at air temp <5°C).*

Figure A3: Automatic Area Fire detector (AAFD) Responses in a Credible Worst Case Fire Test

G-JET Smoke Detection Design Tool – Description

G-JET is a design tool that calculates which category of smoke detector is most suitable and recommends initial nominal sensitivity setting required to detect a given amount of smoke, fully dispersed in a given room or smoke volume. G-JET accepts all categories of smoke detector applications and is independent of product brands. It is based on simple formulae and assumptions. It works within typical national or international prescriptive standards for automatic fire detection, but also complies fully with the referred 'Method 2: Mass Optical Density' for engineering to performance requirements[9]. G-JET lists default values of effective sensitivity (minimum mass of smoke released to be detected), but accepts input values by the user according to stakeholder or other objectives.

G-JET has been in proprietary use for 9 years. It was first published at the AUBE (Automatische Brandendeckung)'99 Conference on fire detection in Duisburg, March 1999[3]. The paper in the book of Proceedings describes the model, definitions, formulae, design tool features and assumptions involved. The paper of the proceedings, an introductory description and a print-out sample is available on the internet at: http://www.interconsult.com/G-jet/.

The core of the model is the equation of optical density relating to mass of material transformed to smoke, fully diluted within a specified smoke volume and to room volume in conjunction with simple, conservative assumptions. Simplicity and usefulness are the prime features of G-JET compared to elaborative design options such as computational fluid dynamics. G-JET calculates the effective sensitivity of any common smoke detection application using aspirating, beam or point type of detectors. Effective sensitivity relates to a mass of given material evenly dispersed as smoke in a room, in a defined smoke volume or in a defined 'cold plume cloud'.

It is a presumption both of the explanatory NFPA "Method 2" and of G-JET that no thermal effects be present - it is the early pre-smokelayering phase, the pre-plume phase, that is being modelled. G-JET models the worst case challenge of smoke detection - that of fully dispersed smoke - as this is typically the most useful design criteria.

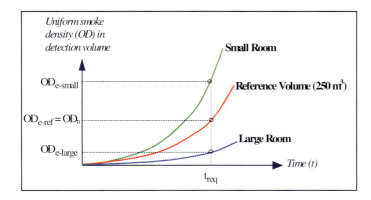

Figure B1: Optical densities in various room volumes as a function of same smoke production

Any other smoke from fires is assumed to layer itself by plumes, typically towards the ceiling where it creates significantly higher optical densities - and at a later phase of fires when damage is more significant. In the plume-phase all listed detectors of nominal (sensor head) sensitivity rated within typical standards will respond reliably within a narrow time frame, thus not necessitating any response modelling for practical purpose.

Sample calculated masses of smoke of PVC required for detectors to respond in various applications.

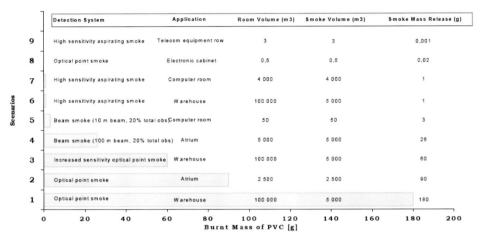

Sample sensitivities of various smoke detector applications, nominal versus effective.

	Office Cell	Broom Closet	Covered Mall	Production Hall (welding)	Computer room (air conditioned)
Room geometry (wxdxh)	3x3.5x2.5	1x1x2.5	80x10x17	50x30x10	5x10x3
Required effective sensitivity S_e	1.5	1.5	5	20	0.05
Part of volume presumed to be occupied by uniform smoke cloud	100 %	100 %	25 %	50 %	100 %
Calculated nominal sensitivity S_n for point smoke detectors	14 %/m	78 %/m*	0.4 %/m	0.7 %/m	0.08 %/m*
Calculated nominal sensitivity S_n for aspirating smoke detectors	1.5 %/m **	14 %/m **	0.09 %/m	0.4 %/m	0.08 %/m
Calculated nominal sensitivity S_n for beam smoke detectors	NA	NA	26 %***	31 %***	0.8 %***
Comparing against conventional practice: Effective sensitivity S_e when applying fixed S_n=1.5 %/m point detectors	0.15 %/m	0.02 %/m	19 %/m	36 %/m	0.9 %/m

*NA = Not Applicable. * = Not Available ** = Applicable by multiroom sensing configuration only. *** Total Light Obscuration*

Table B2: Tables from reference[21] show sample calculation results with G-Jet

Access to G-JET

Enter website http://www.interconsult.com/G-Jet/.
German or English language can be selected in upper left pull-down menu.